Theory of the creation of a 5-dimensional universe
Theory of dark matter and dark energy in a 4-dimensional universe
Theory of black holes

Copyright © 2017 Grzegorz Hoppe
All rights reserved.
ISBN-13: 978-1981938377
ISBN-10: 1981938370

Dr Grzegorz Hoppe

Initial assumptions

1. Let's take the simplest way of thinking, based only on logic and mathematics (as the sole formal sciences) and on the truths (theories) in physics about which we have no doubt whatsoever.

2. Let's not start the existence of our universe from **THAT** (something), the existence of which we cannot (anyhow) explain or understand.

3. The universe is no more complicated than is necessary for its existence (and for everything that we can "see").

What we know for sure

Albert Einstein's special theory of relativity demonstrates that:

$$E^2 = p^2c^2 + m^2c^4;$$

so at rest for p=0, we have $E^2 = m^2c^4$

$$E = \pm mc^2$$

(theoretically, negative energy exists)

All the elementary particles known to us have their counterparts in antimatter (antiparticles), which differ only by the symbol of their electrical charge.

(You could say that they are a mirror image of them)

What we know for sure

Hubble's Law – the basic law of observational cosmology, binding the distances of galaxies D with their escape velocity v (measured by their red shift).

Hubble's law (applicable locally) can also be deduced from general relativity, assuming that the Universe is homogeneous and isotropic. Its expansion is described in the Friedmann equation. In addition to the effect associated with the movement of galaxies, the change in the length of the electromagnetic wave reaching from cosmological distances is also caused by the expansion of space itself.

Therefore, there can be no doubt that **the universe is expanding**.

Hypotheses

Thesis I:
Our observable universe is one of four 4-dimensional universes that together form a 5-dimensional universe. The time dimension is continuous, i.e. there is no t_p – Planck time.

Thesis II:
Dark energy and dark matter are energy and matter belonging to one of the remaining three 4-dimensional universes to which we have no access but whose existence we can experience through the interaction of the energy and matter in our universe with the energy and matter of another. No energy or matter can penetrate the boundary between these universes; therefore, they are undetectable to us.

Thesis III:
Black holes are common objects in space and are essential for the existence of galaxies.

Assumptions

1. The principle (law) of entropy is true

(Entropy is a measure of the degree of disorder in a system and the dissipation of energy. According to the second law of thermodynamics, each isolated system moves towards a state of equilibrium, in which entropy reaches its maximum.)

2. The expression $E = \pm mc^2$ is true

3. Our universe really exists and is real

The beginning

Nothingness (has no definition, is the complement to everything, and therefore cannot exist).

The primordial something must have been **an empty, dimensionless space**.

Using our senses, we can perceive that there is a 4-dimensional universe (three real dimensions plus the dimension of time) which contains matter and energy.

The initial event

Assumption:

In the initial event, two 4-dimensional space-times (space P and anti-space (-P)) were created, with a dimension in common, which is time.

The term 'initial event' may be understood as both the contemporary view of the Big Bang, accepted as a rupture of original space, and also as the act of creation of the Universe. I am of the opinion that this is an unknowable problem and will remain an issue for philosophy, not science.

The initial event – rupture of original space

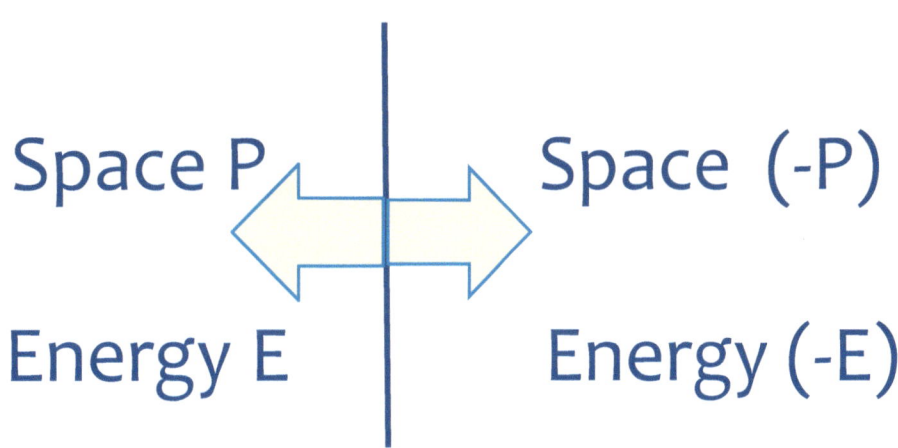

$$P + (-P) = 0, \quad E - (-E) = 0$$

After the initial event

After the initial event, the new space would have been filled with energy, whereby the absolute values of total space energy (E) and total anti-space energy (-E) must have been equal. According to the entropy principle, the distribution of this energy would have had to move towards the lowest possible energy level. The entire universe, however, would have been highly energetic with enough large heterogeneous fragments that it would have had enough time for the creation of all its subsequent particles (elementary), a prerequisite for what we currently observe to be able to exist.

The creation of a 5-dimensional universe

On the basis of the experimentally-confirmed existence of antimatter, which has a mirror structure and mirror properties of matter, we may rightfully conclude that there should be a mirror space in relation to our (-P).

If so, at least two 4-dimensional mirror spaces were created, one filled with matter and energy and the other with antimatter and negative energy.

The creation of a 5-dimensional universe

If, however, only two 4-dimensional spaces were created, they would have to be infinite or limited. If they were infinite (unlimited), then there would be no Hubble's law. Therefore, they are limited, and so a 5-dimensional space was created.

We must therefore assume that a 5-dimensional universe was created. From the principles of mathematics and logic, it must have 4 internal dimensions, from which it follows that there must have been **not two**, but **four**, separate, 4-dimensional spaces.

The creation of a 5-dimensional universe

However, if we consider that these universes must be mirror images of themselves, and that they are all combined into one 5-dimensional universe (and such a combination also necessitates the existence of correspondingly mirrored 3-dimensional spaces), then two identical 4-dimensional universes P and two mirror images (-P) must have been created. This also means that the boundaries connecting the 4-dimensional universes are 3-dimensional universes (remember that the one dimension common to each universe is time).

So the assumptions need to be revised:
Four 4-dimensional universes were created, forming in total a 5-dimensional universe.

The creation of a 5-dimensional universe

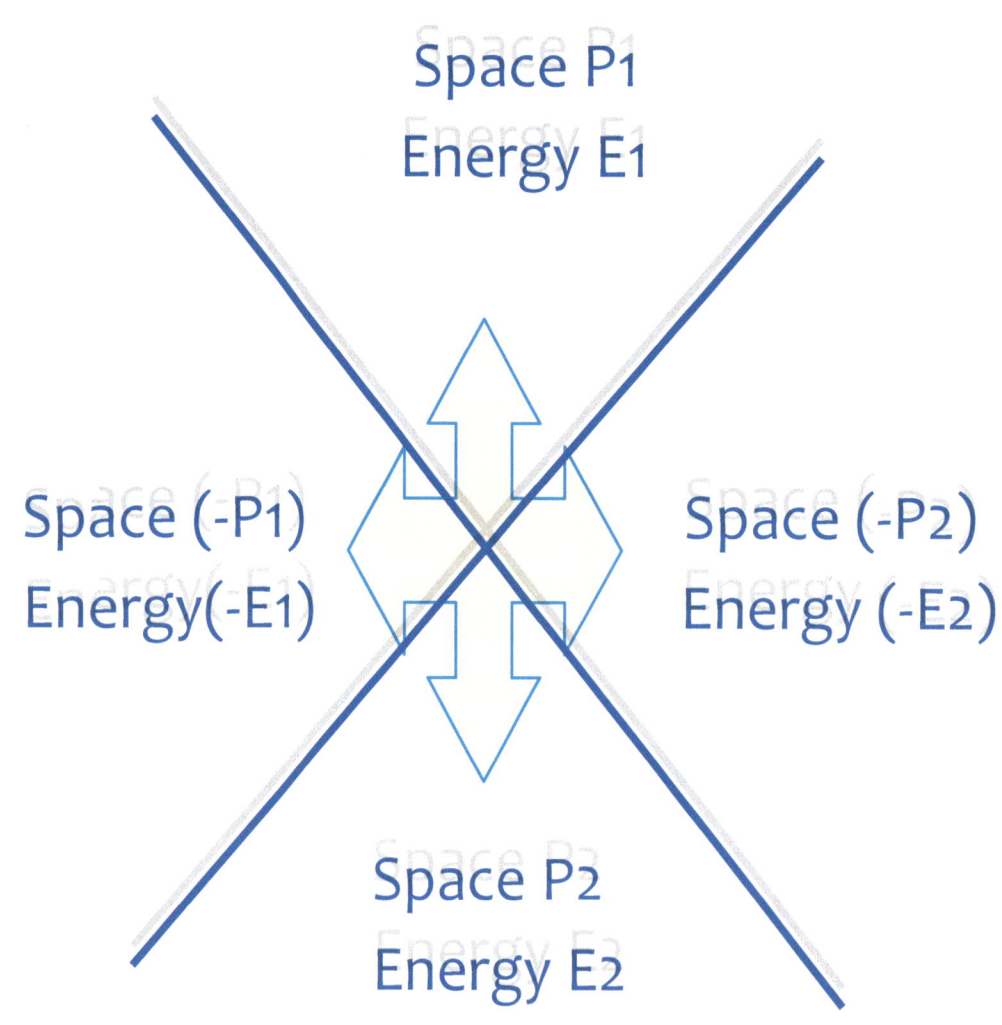

The creation of a 5-dimensional universe

Four 4-dimensional universes immersed in a 5-dimensional universe is an n-sphere with radius r.

In which case, we need to consider whether or not a higher universe i.e. a 6th level, was created.

If a 6-dimensional universe was created, five 5-dimensional universes must also have been created. However, these five 5-dimensional universes would by definition have had to be their mirror reflections creating a 6-dimensional universe. Here, however, we find a contradiction, because the odd number of 5-dimensional universes is at odds with the principle of parity (spatial mirroring). In that case, since there could not have been a 6-dimensional universe with the pre-defined properties of lower-order universes, then, following the rules of logic, there could not also be a universe of an even higher order.

The creation of a 5-dimensional universe

A consequence of this is the necessity to assume that the value of r is infinite or as close to infinity as possible. It also may be assumed that it is finite and limited by nothingness, which is the same, because a 5-dimensional universe can have no limits (beyond nothingness or space of a higher dimension). This also means that r may be constantly increasing in size.

The creation of a 5-dimensional universe

What is a 5-dimensional universe containing four 4-dimensional universes?

It is a geometric object in which all objects of a lower order, and the object itself, have a single point in common. Since, however, the dimension of time is supposed to be the common dimension for all universes simultaneously, this must be the common point. The point, by definition, has no physical dimension (we cannot determine its size), so the **dimension of time** must be **continuous**, not **grainy**, and in such a universe there can be no Planck time t_p.

4-dimensional space in a 5-dimensional universe

According to Carl Friedrich Gauss' *Theorema Egregium*, the property of Gaussian curvatures is their internality, i.e. their differing shape depending from which side we look at them.

This means that the **initial condition for the mirror-like quality** of universes disappears immediately after their creation (following the initial event) because any division of such spaces is a mirror division.

The sole condition for this **initial symmetry** is the uniform (quantitative) distribution of **energy** in all the universes, i.e. each of them should have **25%** of the absolute value **(it is currently estimated that we are "missing" 85%)** of the sum of absolute values for all the universes together (absoluteness due to the positive and negative energy in mirrored universes).

The creation of a 5-dimensional universe (mathematical approach)

$$P_1 + P_2 + (-P_1) + (-P_2) = 0$$

$$E_1 + E_2 + (-E_1) + (-E_2) = 0$$

25 % E of the universe = $E_1 = E_2 = |-E_1| = |-E_2|$

$r \rightarrow \infty$; for the moment t_n ; $r \equiv r_n$

G_1, G_2, G_3, G_4 – boundaries between spaces P

(these are complex planes – Gaussian planes)

The creation of a 5-dimensional universe (mathematical approach)

Dimensions:

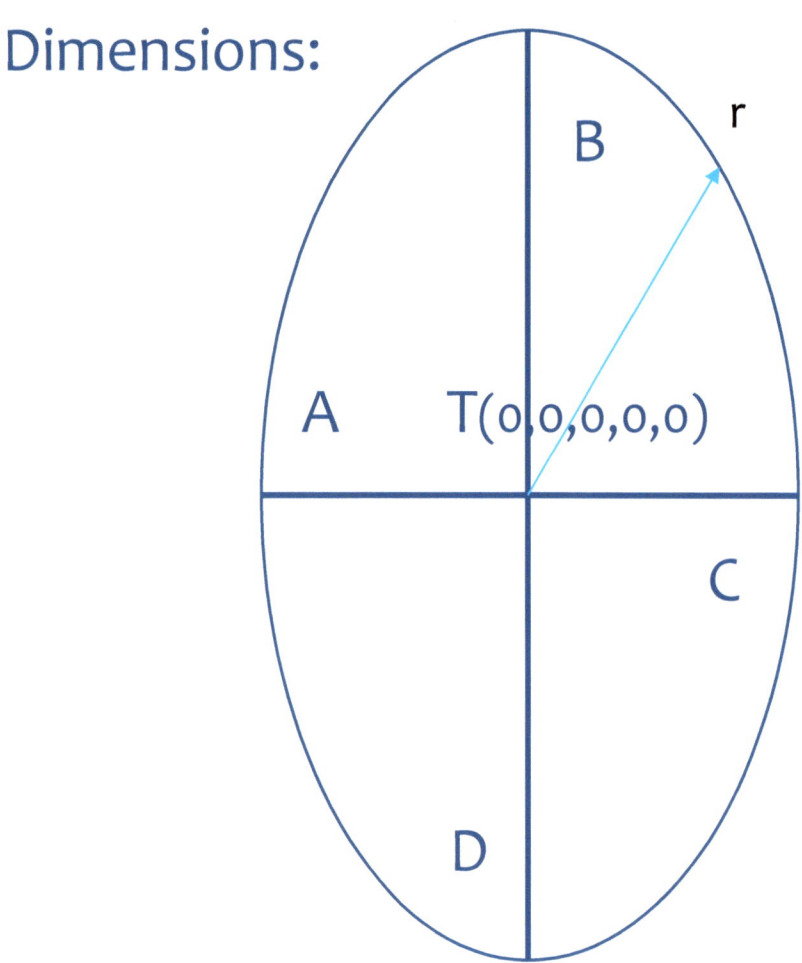

The creation of a 5-dimensional universe (mathematical approach)

$P = (A \times B \times C \times D \times T)$,

for the whole 5-dimensional space
r is a specific value at a given moment
and not a dimension,

$P_1 = (A \times B \times r \times T)$,

$P_2 = (C \times D \times r \times T)$,

$(-P_1) = (B \times C \times r \times T)$,

$(-P_2) = (D \times A \times r \times T)$,

for 4-dimensional spaces r is a vector (dimension)

The creation of a 5-dimensional universe (mathematical approach)

$T(0,0,0,0,0)$ – the centre of all spaces (each dimension),

the only common point in all spaces and the hidden (internal) dimension of time,

The boundaries between 4-dimensional spaces:

$G_1 = (A \times r \times T)$, $G_2 = (C \times r \times T)$,

$G_3 = (B \times r \times T)$, $G_4 = (D \times r \times T)$,

for 3-dimensional spaces (the boundaries)
r is a vector (dimension)

The creation of a 5-dimensional universe

Since we know that the dimension of time is the only common point in all spaces and is a dimension reduced to a point, for the sake of simplifying further considerations the dimension of time can be omitted as a reduced (or internal) dimension and in further considerations accepted that we are dealing with a 4-dimensional universe consisting of four 3-dimensional universes.

Geometrically, such an object will be an n-sphere (torus) with radius r, and its internal walls (separating the 3-dimensional universes) will be Gaussian curves (planes). Thus, within each of the 3-dimensional spaces, we have a Gaussian curve (plane) on the walls separating the 3-dimensional spaces from each other. In addition, if r = ∞, then we will have Riemann spheres (a complex closed plane). If, on the other hand, r is approaching ∞ then we are dealing with **cosmic inflation of space** (a proven fact).

5-dimensional space
P (A x B x C x D x T)

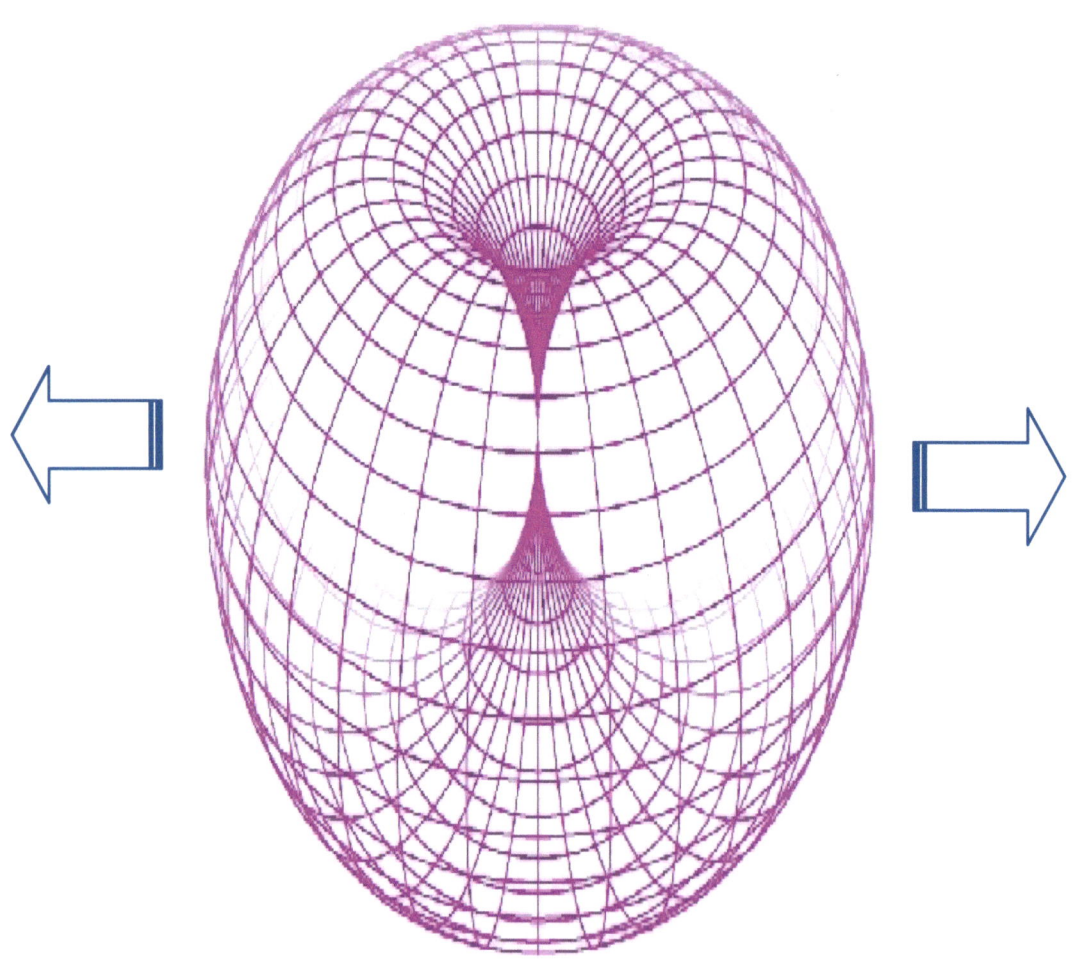

3-dimensional spaces (boundaries) in a 5-dimensional universe – Gaussian planes

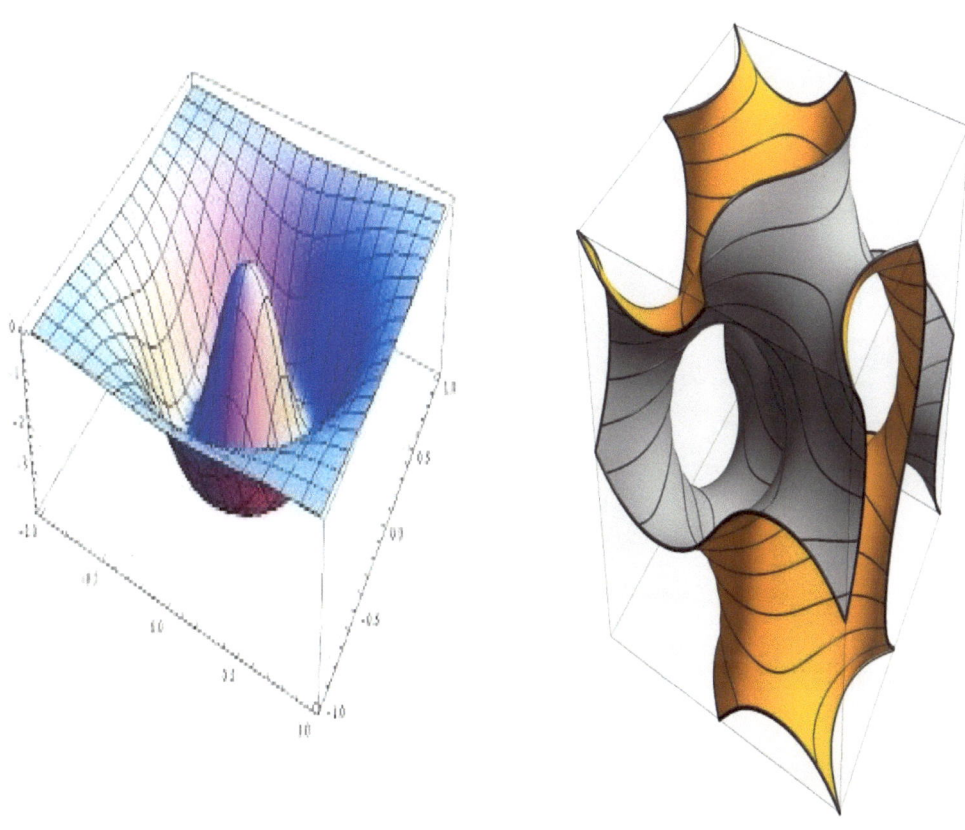

3-dimensional spaces (boundaries) in the 5-dimensional universe – Gaussian planes

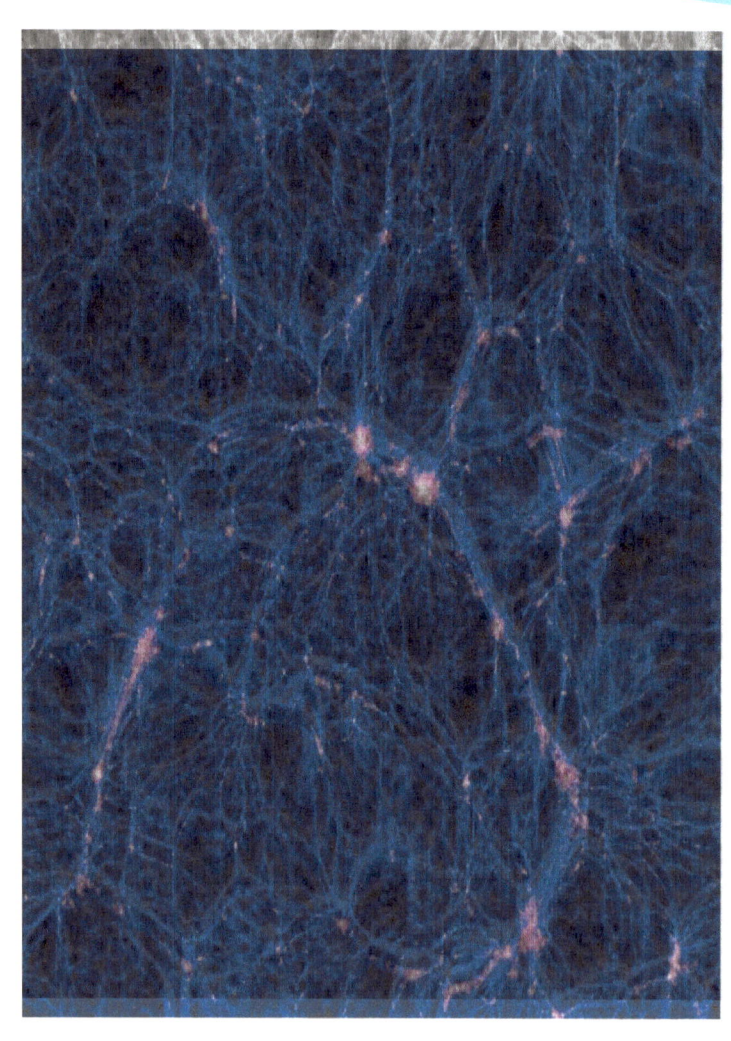

Dark matter and dark energy

If we wanted to transfer this description to our reality, it would mean that the boundary with the remaining universes could have any possible shape, and our universe should therefore have 25% of the total value of all existing energy (in the absolute dimension).

It is also worth noting the important fact that in such a universe the laws of complex numbers (imaginary) for calculations (taking into account its boundary) would apply.

As was previously demonstrated, the boundaries between these individual universes are Gaussian planes, which can be any shape for an observer from within one of the universes. Their boundaries must be constructed of planes that do not let any energy or matter pass through, otherwise all the universes would have been annihilated long ago.

Dark matter and dark energy

If objects limited by such boundaries were in our universe, then, for us, they would be invisible planes through which the matter and energy of our universe could move as if they did not exist (never penetrating inside such an object).

If we assume that the value of radius r in the n-sphere described is constantly increasing, we can explain the expansion of our universe, as has been observed.

All contemporary knowledge of dark matter and dark energy is consistent with the above description including, above all, the inability to observe either, along with the estimated amount of matter and energy in our universe and its total size. In the light of these, and the previously confirmed facts, it must be therefore accepted that the assumption (2P + 2 (-P)) is correct.

An atom is 99% empty space

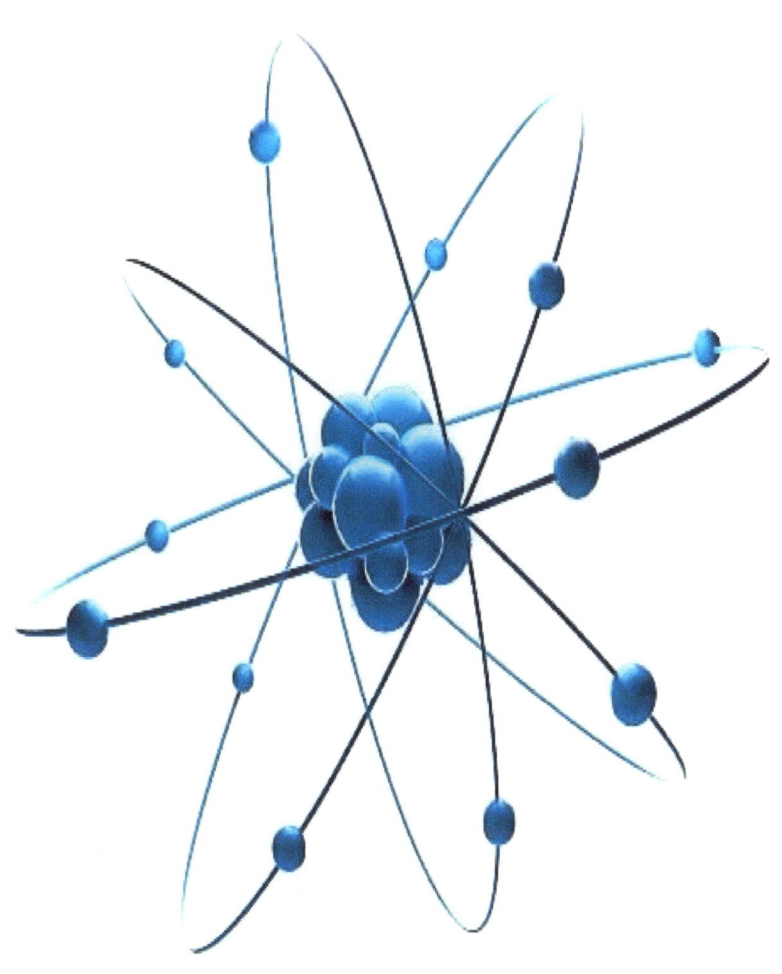

Black holes

Black holes arise from the formation of extremely heavy, extremely hot, objects in space. Because the structure of the matter is almost empty, this matter becomes increasingly compressed in a given space. The process ends in the collapse of the matter's particles, causing the existing spaces between its elements to be filled only with subatomic particles. As a result of this process, the temperature of such an object increases constantly (its energy increases). At the precise moment when the temperature increase of the object reaches its maximum (critical) value, the object collapses and turns into a black hole.

A black hole – a sphere with the maximum possible level of energy

Black holes

A black hole is a spherical object which spins around its own rotation axis and consists of particles of matter with their maximum possible energy, and thus their maximum temperature. Photons colliding with such a sphere are unable to pass through it because they can no longer increase the sphere's energy. For this reason, photons increase the energy in the external matter surrounding this spherical object, in which there are still elements of matter that can increase their energy, and which they "sink into", releasing their energy. Due to the fact that the level of energy in the interior of the black hole is the maximum possible, no energy wave that could be observed from the outside of the black hole as its radiation can escape from it. As a result of this process, the object reflects no waves and becomes invisible to us or, more precisely, perfectly black (which is consistent with observations).

Black holes

The impact of a black hole on photons running close to it can be observed as a curvature in their path. Because the natural path of photons is the shortest possible (which is simple in the absence of external influences), where they approach such a massive, hot object their path must be curved, since this is the shortest possible path (the light curvature effect).

The outer structure next to the black hole appears as a vortex of matter and energy (photons), (**accretion disk**) which for the above reason slowly become part of every black hole (they are absorbed). This applies to that part of the matter separated from the black hole by a certain distance, which is the magnitude of its impact. It is obvious that the larger the black hole, the larger the radius of its impact.

Black holes

Due to the way a black hole is constructed, i.e. as a sphere of matter sucking in matter of lower energy to its interior in a vortex motion, it has its own axis of rotation. The matter rotating around a black hole collects at its "equator" (and maintains its temperature – the maximum energy of a black hole).

At its poles, vortexes form which can cause the surface of the black hole to come into contact with the external cold space, which leads to an ejection of energy and matter from the black hole (**jets**). Where part of a black hole's energy is ejected, it is directed towards one of the axes (one of two, because of course it exists on both sides). The diameter of the beam of ejected energy and its range depend on the size of the surface of the black hole that has come into contact with the external cold space. This happens in accordance with the principle of entropy (which is why we observe different jets – both small and powerful).

A black hole – accretion disk and small ejection of energy

A black hole – accretion disk and large ejection of energy (jet)

Black holes

Black holes are objects in space which are necessary for any galaxy to exist. At the centre of every galaxy there must be the most massive black hole, which forms the galaxy's axis of rotation. In each galaxy, depending on its size, there may be any (finite) number of black holes that form a structure holding the galaxy together. This is a prerequisite for the counterweight to the entropy principle (**cosmic inflation is suppressed**). Otherwise, galactic matter would get scattered about and the galaxy would be destroyed. All black holes in a given galaxy interact with each other to create the above mentioned structure. In the case of a black hole with a mass (energy) greater than the most massive in a given galaxy, the galaxy's axis changes, so that it is always the most massive black hole (everything is consistent with the observations).

The essence of how our universe works

The observable fact of the movement of all galaxies implies that the most massive black hole in our universe must be its centre and the axis of rotation of all existing matter and energy, each of the four 4-dimensional universes (**in the middle with point T(0,0,0,0,0), hidden, internal dimension of time**).

The theory described in this way is consistent with the "reality" we can observe, so the assumptions must be correct. A very important consequence of this theory is the need to accept that **time** and **space** are **continuous** and not **grainy**.

The essence of how our universe works

If we accept that the rapid expansion of original space is a **natural property of empty space**, then we can logically explain observed **quantum fluctuations** and the entirety of quantum mechanics.

Because such expansions are obviously limited (in terms of size) by the existing structure of the universe, they can result in at most four elementary particles or energy quanta, which in total have zero energy and zero mass.

This is our universe

This theory rests entirely on the principles of formal logic and mathematics.

Its only assumptions were the acceptance of the truth of the entropy principle and the truth of the identity of energy and mass ($E= \pm mc^2$).

As a result, the theory must be true and must describe our reality. If another theory should be contradictory, this means that it is false.

Unless it isn't our universe

If we believe at all that the universe had any beginning, (the first event - the initial expansion of space, the Big Bang or the act of creation), which led to what we observe (that is, we reject the hypothesis that our universe has always existed), then the 5-dimensional universe I have described must have been created anyway, unless my theory has a gap in logic.

If it must have been created, it must exist. The question is only whether it is our universe or not (or are we living in a different design of universe). In that case, we need to think about what speaks for the fact that this is our universe (have we not observed enough facts yet?).

So if it was not actually our universe, we would then also have to accept that there were at least two initial events (different). Can there be two different initial events?

It can't there be two different initial events

Dr Grzegorz Hoppe, December 17, 2017

www.ingramcontent.com/pod-product-compliance
Lightning Source LLC
Chambersburg PA
CBHW051931210526
45473CB00006B/2211